A Guide to the Preparation of County Road Histories

Virginia Genealogical Society
Richmond, Virginia

Published With Permission from the

Virginia Transportation Research Council
(A Cooperative Organization Sponsored Jointly by the Virginia
Department of Transportation and
the University of Virginia)

HERITAGE BOOKS
2008

HERITAGE BOOKS
AN IMPRINT OF HERITAGE BOOKS, INC.

Books, CDs, and more—Worldwide

For our listing of thousands of titles see our website
at
www.HeritageBooks.com

Published 2008 by
HERITAGE BOOKS, INC.
Publishing Division
100 Railroad Avenue #104
Westminster, Maryland 21157

International Standard Book Number: 978-0-7884-3367-2

A GUIDE
TO THE
PREPARATION OF COUNTY ROAD HISTORIES

By

Nathaniel Mason Pawlett
Faculty Research Historian

Virginia Highway & Transportation Research Council
(A Cooperative Organization Sponsored Jointly by the Virginia Department
of Highways & Transportation and the University of Virginia)

Charlottesville, Virginia
March 1979
VHTRC 79-R12

PREFACE

A Guide to the Preparation of County Road Histories is the tenth publication of the continuing series "Historic Roads of Virginia" issued by the Virginia Highway and Transportation Research Council. It is intended to set out in simple, easily understood terms the knowledge gleaned from the research and writing of Albemarle County Roads 1725-1816 and the associated volumes of this series; the sources of information used, as well as other possible sources; the information recording techniques used; certain useful rules of evidence, as well as some of the pitfalls to be avoided; how the analysis of this information proceeded; and, finally, and probably most importantly, an attempted formulation of some minimum standards for an acceptable road history. Although the experience of the more western areas of the United States in setting up their road systems will be found to be somewhat different from that of those with systems deriving directly from English precedent and the basic English road law of 1555, this guide, perhaps with some slight modification in techniques, should have broad general applicability.

A GUIDE TO THE PREPARATION OF COUNTY ROAD HISTORIES

By

Nathaniel Mason Pawlett
Faculty Research Historian

INTRODUCTION

The importance of the preparation of a detailed history of the early roads of each county can hardly be overstated. Most of this early road network still remains in place and in service. With subsequent, sometimes extremely subtile, changes in emphasis, and with a few additions here and there, it has over the years provided the lines along which the Commonwealth's social, political and economic life has flowed. Indeed, the fate of the nation has at times been determined by this road network. Beyond this, road history and its supporting documents are becoming relevant to a growing number of disciplines from anthropology and architectural history through environmental science, folklore, genealogy, and history to sociology.

Local historians and genealogists, perhaps the most obvious groups to which these publications have relevance, have also a vested interest in seeing that more of these road histories are completed. Moreover, these are the people with the specialised knowledge of local history and genealogy so necessary to the writing of a road history. If in our time this massive picture puzzle whose pieces still lie before us in a heap is to be reassembled, an effort must soon be made. This present modest effort is the pioneering one, designed only to begin the task by attempting to sketch in broad outlines those roads laid down during the first 50 years after settlement in the area presently contained in each of Virginia's counties. That this must be done by local people, or those with a similar orientation and knowledge, should be obvious. That it must be done in this generation, before the rapidly quickening pace of urban and suburban development obliterates much of the visual evidence, is manifest. Local historians and genealogists are therefore summoned to a duty which has perhaps already been too long delayed in many areas.

The Virginia Highway and Transportation Research Council is a cooperative organisation sponsored jointly by the Virginia Department of Highways and Transportation and the University of Virginia and is located on the Grounds of the University at Charlottesville. The Council engages in a comprehensive program of research in the field of transportation. As a part of its program the Council, in December 1972, began research on the history of the road and bridge building technology in Virginia. The initial effort was concerned with truss bridges; a complementary effort concentrating on the early roads of Albemarle County got under way in October 1973.

The evolution of the county road systems of Virginia is in many ways inseparable from the social, political and technological developments that form the history of the Commonwealth. Despite this, few serious works exist on the history of roads in Virginia. Those which have been produced tend to focus on internal improvements and turnpike development at the state level before the War Between the States. Little has been done on the period from Reconstruction through the creation of the system of state highways in the earlier part of this century, while the early road systems of the individual counties have at most received treatment of from a few paragraphs to several pages in local histories. Accordingly, it was decided to investigate the development of the roads of Albemarle County during the period 1725-1816 as a pilot project, and, using this experience, to produce a history of Albemarle County roads and a handbook on procedure to aid others in writing similar road histories.

Before continuing further it might be helpful to briefly summarise the English system of road administration as it evolved in Virginia. The establishment and maintenance of public roads were important functions of the County Court during most of the colonial period in Virginia. Each road was opened and maintained by an Overseer or Surveyor of the Roads charged with this responsibility and appointed by the Gentlemen Justices. He was usually assigned for this purpose all of the "Labouring Male Titheables" living on or near the road. Major projects, such as bridges over rivers, demanding considerable expenditures were usually executed by Commissioners appointed by the Court to select the site and contract with workmen for the construction. Where bridges connected two counties, a commission was appointed by each and they cooperated in executing the work. Each order for a road, as well as those for bridges, was entered by the Clerk in the order book of the County Court.

During the early stages of the Albemarle project it was necessary to examine and extract all the road orders for the counties from which Albemarle was formed as well as the Orders for Albemarle when it still contained the Counties of Amherst, Buckingham, Fluvanna, Nelson, and a part of Appomattox. The broad applicability of those orders from Goochland, Louisa and early Albemarle, as well as the opinions of various authorities throughout the state who examined them, indicated that they should have separate publication in order to make them generally available to individual scholars' through libraries and educational institutions. These orders made up the first three publications in the series "Historic Roads of Virginia".

In December 1974, while these were being prepared, a request was received from James A. Bear, Jr., Resident Director and Curator of the Thomas Jefferson Memorial Foundation at Monticello. Mr. Bear, engaged in editing Thomas Jefferson's account books for publication, wanted to know the original route of the Three Notch'd Road between Richmond and the Valley.

The report prepared for him elicited so much interest that it was gradually expanded to include a folding map, the results of a reconnaissance by two Council members, an appendix of the pertinent information from the Order Books of Louisa, Goochland, Albemarle and Orange Counties, citations in the <u>Virginia Gazette</u> of Williamsburg, and photographs of portions of the road, abandoned and still in service. Accordingly, it was determined to issue this accumulated information as the fourth publication of this series.

During the research at Albemarle County Court House three Surveyors Books covering the years 1744-1853 were discovered. The plats contained in these books showed many roads in present Albemarle County as well as Amherst, Buckingham, Fluvanna, Nelson and parts of Appomattox, Bedford, and Campbell Counties, which were at one time within it. Photographic copies were made of these plats and, after annotation, an index was arranged to provide for easy reference as the remaining road orders of Albemarle County (1783-1816) were being analysed and indexed. As the utility of this index came to be recognised it was decided to issue it as another publication in this series (the fifth).

Experience with the Three Notch'd Road had by now clearly demonstrated the feasibility of preparing brief reports setting out the specific route of a road and the principal facts concerning it. From road orders available at the county courthouses, published Acts of the Assembly, the records of the Board of Public Works, Confederate Engineers' maps and those published by the Virginia Department of Highways and Transportation, similar brief histories and route maps could easily be constructed for many of the eighteenth and nineteenth century roads of Virginia.

Especially was this true for turnpikes. By this time the records of the Board of Public Works 1816-1902, Virginia's coordinating body for internal improvements such as canals, railroads and turnpikes, were available on microfilm, thanks to the efforts of Howard Newlon, Jr. of the Council and Donald Haynes of the Virginia State Library. Most of the Confederate maps were also available at either the State Library or the Virginia Historical Society in Richmond, Architectural surveys such as those available at the Virginia Historic Landmarks Commission and the School of Architecture at the University of Virginia often provide additional information as to toll-house and tavern locations along the routes. Besides this, most of the turnpike routes are still in service as primary or secondary roads, thus simplifying greatly the task of the amateur road historian. With the foregoing in mind, a brief history of the Staunton and James River Turnpike became the next publication of the series. Originally a senior undergraduate thesis in the humanities program of the School of Engineering and Applied Science at the University of Virginia, this publication was issued in the belief that it was a good example of what individuals in local historical groups could accomplish in a limited time with the use of readily available resources dealing with the hundreds of Virginia turnpikes of the eighteenth and nineteenth centuries.

Following this, the long-awaited and massive volume (421 pages) <u>Albemarle County Road Orders 1783-1816</u> was indexed and published and the Albemarle road history itself was begun. Since no adequate treatment of Virginia's roads up to the advent of the steam railroad was available, the first portion of the Albemarle study was given over to a synopsis of this from about 1607 to 1840. At the suggestion of several of the people who read the first draft of this, and in the interest of increasing the utility of this particular section, separate publication was decided upon and A Brief History of the Roads of Virginia 1607-1840 came out in the fall of 1977, followed by Albemarle County Roads 1725-1816 in 1978.

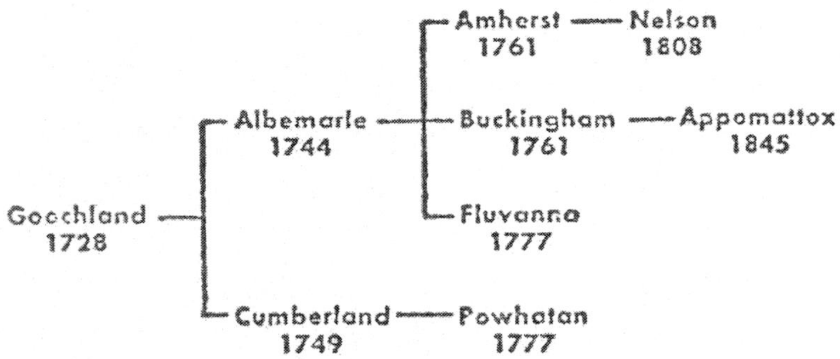

Figure 1. A chart illustrating the evolution of Albemarle County.

PREPARING A COUNTY ROAD HISTORY

The Provenance Chart

The primary source of information on the roads of a county is the road orders issued by the gentlemen justices of the county court and recorded in the order books of the county. Directed to one specific individual, usually styled overseer, or surveyor, of highways or roads, these orders briefly described the road or portion of it to be built or maintained and set forth in a very general way those labouring male titheables who would maintain it. Usually issued yearly with the appointment of new surveyors, these orders constitute the bedrock of the foundation of any road history worthy of the name. Without them, certainly without the major part surviving, it would be impossible to speak with any degree of authority concerning the development of the roads of a county. Although surviving vestry books, deed books, and the records of adjacent counties might prove of some benefit, the road orders are essential. All other records, as well as maps, provide only supporting evidence to that of the road orders.

Since the geographic boundaries of a county have likely changed through the years as other counties were created from it, a provenance chart for the county under consideration should be prepared as a preliminary to any historic survey of a road or roads. This should show the pilot county, its date of formation, whence (and when) came its component parts, their derivation back at least to the threshold of settlement of the area in the pilot county, the provenance of the adjacent counties, their later progeny, etc. This chart is designed to enable the investigator to locate the records of the areas within the pilot county at any given point in its development. All additions to or subtractions from the areas of the counties should also be entered on the chart, which is best prepared in the form of a block diagram. Figure 1 is an example of the most basic form of this diagram, with the development of adjacent counties yet to be added. The importance of this chart cannot be overemphasised. It might also be noted here that county lines in some cases, such as where they follow watersheds, tend to be rather indistinct and that the date of formation, subdivision, etc. of a county cannot be relied upon as an absolute index of where records will be located. On occasion an overlap of several years may occur after formation of a county before all its records consistently occur within it.

The first step in the writing of a county road history, or an individual road or turnpike study, should be the construction of a comprehensive provenance chart similar to that for Albemarle County shown in Figure 2. This chart illustrates graphically the development of Albemarle County, its antecedents, and the adjacent counties, and the condition of their records today, with the primary emphasis remaining on the records of Albemarle County. Its basic form was constructed from charts (see Figure 1) found in Morgan P. Robinson's <u>Virginia Counties: Those Resulting from Virginia Legislation.</u> The experience gained in the study of Albemarle roads will be used to illustrate the usefulness of this chart. Since some roads obviously existed in the area of Albemarle County at its formation in 1744, the chart, once prepared, indicated that the starting point for research lay somewhere within the records of its antecedent counties at that time when the first road being opened westward approached its borders. As the earliest land patents in the area were known to date from around 1730 it seemed logical to expect to find roads then beginning to penetrate the area as a result of the settlement of these. To be on the safe side Henrico County records were first examined from about 1720 to the creation of Goochland County in 1728, but this effort was without result. Goochland County records were more productive, showing the first road being built into the area about 1731. Since this was so close to its date of formation (1728), it was decided to record all the Goochland road orders from that date forward to the formation of Albemarle in 1744.

In addition to Goochland, which contributed approximately the lower two-thirds of the present Albemarle County in 1744, Louisa also played a significant part in the development of the county. In 1761, as a result of the subdivision into several counties of the Greater Albemarle of 1744, what was then the western portion of Louisa County was added to Albemarle. This meant that the records of Louisa and its parent county, Hanover, had also to be searched. As the provenance chart so clearly reveals, almost all the early records of Hanover are lost, as well as the Louisa order books from 1748 to 1766. Therefore, only those orders in the surviving first order book of Louisa could be recorded. Although it is regrettable that those for this area in the period 1730-1742 are not available, it was possible with the aid of the surviving 1742-1748 orders, and using the experience of that adjacent portion of Goochland during the same period, to draw some conclusions regarding road development in the area. These are probably accurate within a range of two or three years either way.

Besides enabling those doing research to know in which county or counties the area within the present county lay at each period of its development, once prepared the chart shows at a glance the condition of their records, if their coverage is discontinuous, etc., and it can, if necessary, be expanded to show where such information is presently located if it has been moved from the county to such repositories as the Virginia State Library in Richmond. For counties now in Kentucky and West Virginia, important records may be found here and in the counties from which they were originally divided.

Since a good provenance chart will also show the parallel development of the adjacent counties, their parent counties, and additions and subtractions which have been made to their areas, it should prove invaluable in cases where it is necessary to consult the records of one of these counties. The Albemarle chart here provides a case in point. During the preparation of the Three Notch'd Road report it became necessary to ascertain the date of the first road through Wood's Gap in the Blue Ridge. The chart indicated that the records for Augusta County were, after the formation of the county in 1738 and prior to a county government being set up there in 1745, located in Orange County whence it was governed during this seven-year period. An examination of the order books of Orange County produced orders for this road in 1738 or 1739 along with marginal notations by an early historian indicating that the entry dealt with Wood's Gap rather than Rockfish. Had the chart

PROVENANCE CHART

ALBEMARLE COUNTY AND ITS ENVIRONS

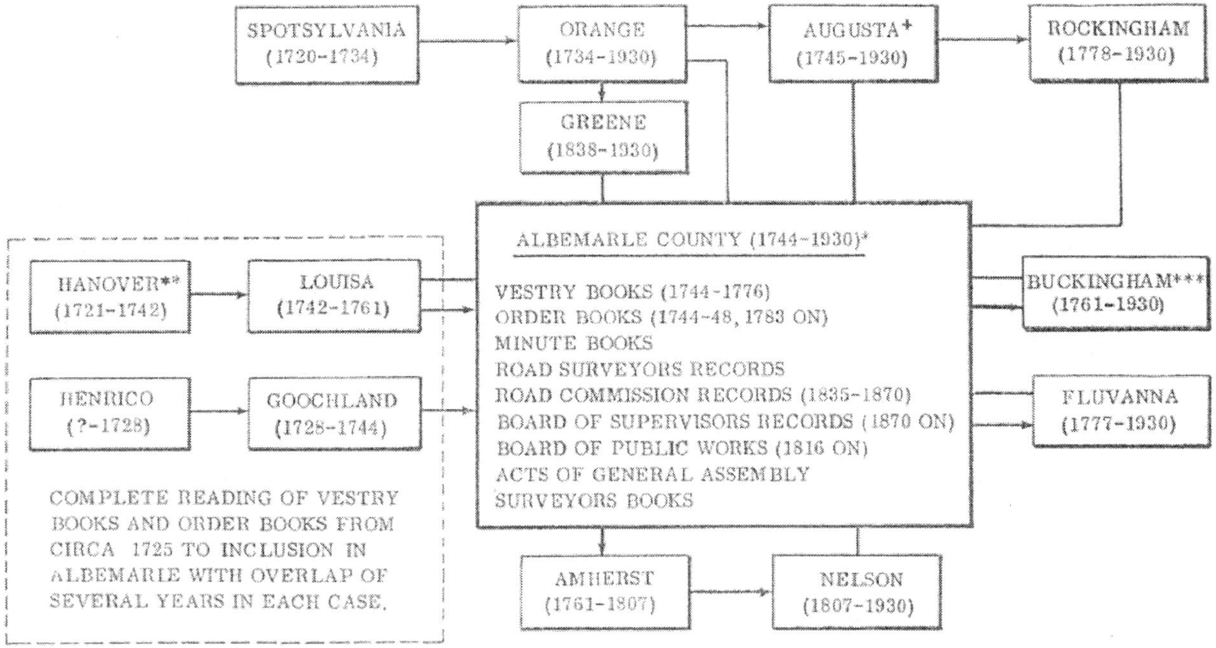

CODE: ARROWS SHOW DERIVATION OF COUNTIES.

SOLID CONNECTING LINES INDICATE A CURRENTLY ADJACENT COUNTY.

\+ RECORDS FOR THIS COUNTY PROBABLY IN ORANGE UNTIL AFTER 1745.

* ORDER BOOKS DESTROYED (1748 - 1783)

** DESTROYED OR LOST WITH EXCEPTION OF SMALL BOOKS (1734 - 1735)

*** BURNED 1869

not been prepared a useless trip to the courthouse of Augusta County might have been made.

The Development Map

The second step preparatory to the writing of a county road history, an outgrowth of and a useful adjunct to the provenance chart, is the construction of a developmental map or maps of some sort. This may be only an elemental county-outline map of the form shown in Figure 3, a larger composite, or a set of them, made up of several county road maps, or an even larger set of geological survey maps. It is even possible to conceive of a situation in which all three might be necessary in order to gain an adequate understanding of the development of a particular county or tier of counties somewhere in Virginia.

While the provenance chart for a county can usually be constructed without undue difficulty using the Acts of Assembly and the various books on the formation of Virginia's counties, drawing the developmental map or maps can become a much more complicated task. This can occur because of the difficulty in ascertaining the exact location of county lines at a given period, if through later separation or evolution of the county they were altered. Although defining the county line created no real problem in the Albemarle project, our experience might be cited here as an example of the sort of thing which will often occur.

That the western portion of Louisa was incorporated into Albemarle in 1761 is a fact known and cited by both authors and authoresses of many of the articles and books written over the years on the history of Albemarle County. The exact location of the line erased by this amalgamation, as well as the magnitude of that portion of Louisa amalgamated, remained" a matter of conjecture until very recently. Even the most recent full-length history of Albemarle County, published in 1976 (John H. Moore, Albemarle Jefferson's County 1787-1976) gave no indication of the true size of the area given to Albemarle by Louisa. Other authors earlier had indicated that there had been no county line in this area or that it had remained indefinite up to the year 1761.

With the recent publication of abstracts of Louisa deeds 1742-1774 all of these statements were invalidated. (Louisa County, Virginia, Deed Books A and B 1742-1759, Louisa County, Virginia, Deed Books C, C 1/2, D and D 1/2 1759-1774, by Rosalie Edith Davis). These abstracts showed quite clearly that a line had existed prior to 1761, and that it ran straight in what might appear to be a rather illogical manner, crossing and recrossing a river so as to cause certain areas one would normally expect to fall within the adjacent county to fall within Louisa.

Fortunately, as stated, this was an interior line destroyed by the alterations of 1761. Had such been the case with one of the lines constituting the present outer boundary of Albemarle County, drawing an accurate developmental map would have been made considerably more difficult, and would have involved research in the deeds of Albemarle and the county adjacent on that side. As it was, a fairly accurate developmental map of Albemarle could be sketched, the principal error contained in the first effort being the omission of that portion of early Albemarle lying to the south of the James River from the head of the Appomattox River to the Blue Ridge.

Surveying the Sources

Once the basic provenance chart and a developmental map have been prepared, a detailed survey of the information sources still available to the researchers can be begun. This can, and probably should be, as detailed as possible, and list all the records available for the subject county. The survey should be accomplished by as many visits as are necessary to the county courthouse, the Virginia State Library, and any other repositories of relevant information. All of these sources should be listed on the provenance chart_ Details concerning the condition of the records, discontinuous coverage or whatever else is of importance may also be entered.

Although Figure 2 is detailed for only Albemarle County, more serious students of road history will probably wish at this stage to visit the antecedent or parent counties, and those counties adjacent, and to enter on the chart similar information regarding them since it may also ultimately prove necessary to make a detailed examination of the records of those counties. Obviously a provenance chart can be as complete or as elemental as the researcher wishes to make it, but it is wise to construct a fairly comprehensive document at the survey stage of the project. Much time will be saved as the work progresses if a copy of this chart is placed in the hands of each of the researchers and kept always at hand. Just what this chart will contain when completed will probably vary from county to county according to the completeness and the condition of their records, as well as the time period the study will cover. Order books, minute books, road surveyors records, road commission records, and surveyors books could be enumerated, along with the printed Acts of Assembly and those of the Board of Public Works. Vestry books are valuable sources down to the time of the disestablishment of the church, at which time the vestries lost their character as quasi-political bodies. Whether available in printed form, on microfilm at the Virginia State Library, or occasionally still residing at the county clerk's office or elsewhere locally, they should also be entered on the provenance chart.

Beyond those sources roughly categorised as public records, a bibliography consisting of county histories, memoirs, and related works dealing with the area might be compiled and kept with the provenance chart, perhaps on the back of each copy for ready reference. Such early maps as exist for the area should also be listed and copies secured if possible from the holding agencies such as the Virginia Historical Society, Virginia State Library, and the Library of Congress. If this is not possible, then the holding agency or the nearest location where a copy can be consulted should be listed. Land Ownership Maps [Library of Congress, 1967] will list most (but probably not all) of the Virginia county maps of the nineteenth century, including those done by the Confederate Engineers in 1863 and 1864. (See Figure 7).

Hanover and Goochland Counties, *circa* 1740
Superimposed on a Modern Map

Louisa and Goochland Counties, *circa* 1742

Louisa and Albemarle Counties, *circa* 1744

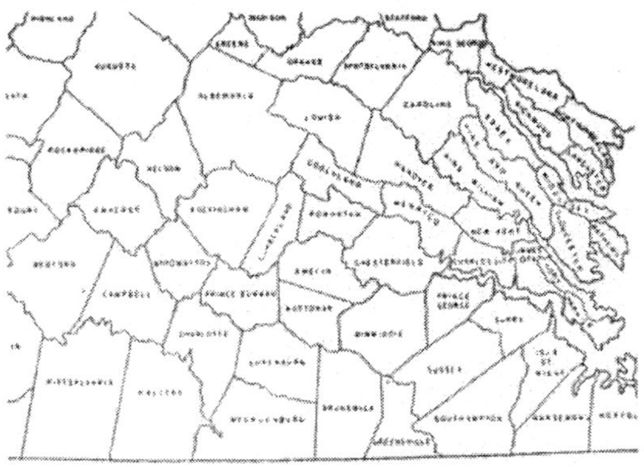

Albemarle County, *circa* 1762

Figure 3.
Development of Albemarle County

Although this publication appears to be comprehensive, and is a good starting point, certain maps do exist which are not listed therein. Maps, much like gold, are where one finds them.

All of the foregoing information from the survey of sources should fit rather easily onto the two sides of a legal sized sheet of paper without undue crowding or the necessity for miniaturisation. If it is desired to provide similar comprehensive listings for the adjacent counties, a separate sheet may be resorted to for each of those.

There are a number of additional sources which, while probably not worthy of inclusion within-the provenance chart, should be included in this survey. Newspapers, particularly those of the nineteenth and early twentieth centuries, often can serve to confirm or to elucidate what is either known or suspected concerning roads and bridges. Not so useful in the eighteenth because of the smaller number of papers then publishing, they may in specific instances still be used to some advantage. Surviving papers and their locations are listed in Lester Coppon's Virginia Newspaper. 1821-1935 (New York, 1936).

Considering the fact that many or most of the road orders define the areas of road under consideration by reference to individuals, their plantations and mills, local genealogists will become a necessary adjunct to the road research, a relationship which it is hoped will prove mutually beneficial. This hope will be at least partially realised if, as in the Albemarle project, the accumulated road orders are indexed and published. Of course, the best arrangement of all would be for those doing the research itself to be people already possessed of a strong inclination toward local history and genealogy, and, most importantly, some-experience working in these fields.

Such well-known statewide resources as E.G. Swem's Virginia Historical Index, the Virginia Magazine of History and Biography, the William & Mary Quarterly and others may prove useful, along with the manuscript and microfilm collections at such major repositories as the Virginia State Library, where most extant county court records up to 1865 are available on film. Also useful is the listing of county histories in Virginia Local History (Virginia State Library, 1971, 1976 supplement), although the spate of new county histories recently issuing from the various local groups tends to make this somewhat out of date.

To sum up this section on the survey of the sources for a history of the roads of a county, the author would like to emphasise the importance of those official records dealing directly with roads themselves or, as in the case of plats and maps, illustrating their locations. These will be the primary source materials of the study, while those others which serve to shed some light on one or another of the many facets of roads and their development can be only secondary or supporting evidence. The question of what constitutes valid evidence will be dealt with and some rules drawn from the author's experience set forth in a later section. Before this, it is perhaps in order to look at some of the techniques used during the Albemarle study for the recording of evidence of various forms and for its correlation with maps, present roads and other forms of evidence such as surviving road traces.

<u>Recording the Evidence</u>

The road and bridge orders contained in the order books of the counties from which Albemarle ultimately came to be formed were a primary source of information for that study. Those from 1728-1816 were extracted and ultimately indexed and printed by the Virginia Highway and Transportation Research Council. All of the county court order books used were in manuscript, often damaged and faded so as to be almost indecipherable. Some of the earlier eighteenth century materials were in the rather ornate, engrossing script and copperplate of the time (see Figure 4), with the phonetic spellings of this period often serving to further complicate matters for the researcher.

In order to facilitate the gathering of this material and its exact reproduction, a system was devised to render these road orders <u>literatim</u> into a hand-held dictaphone tape recorder. This system might be set out in the following rather general rules:

1. Capitalisation to be so stated. Viz: "cap Wadlow cap Cuthbert cap Twaddle" in the case of the name Wadlow Cuthbert Twaddle and any other words in the road order which are capitalised.

2. Names with variant or phonetic spellings to be spelled out.

3. No periods unless so stated.

4. Periods, commas, colons, Semicolons, etc., to be stated.

5. Date and pagination in order book, vestry book, <u>etc.,</u> to be stated. Pre-1752 dates to be designated O.S. for Old Style.

6. Each new citation to be so stated.

7. No paragraphing unless so stated.

8. Tape reels to be marked sequentially as completed with county, name of record, book number(s), and approximate date covered.

Following this, the orders were put into typescript by a series of patient, long-suffering secretaries who had to unlearn all the modern rules of spelling and then learn to render the orders exactly as dictated into the tape recorder. Following this transcription, the tapes were compared to the transcripts by the original recorder and errors were corrected. After being placed in their final form and indexed, the road orders were ready for publication. In published form the road orders were then distributed to a wide variety of organisations and individuals engaged in historical and genealogical research, as well as sociologists, folklorists, and people in a variety of other disciplines.

With these road orders available in printed form, it was possible for an individual using a copying machine, scissors, and paste pot (or stapler), to produce chronological chains of road orders illustrating the development of each of the early roads from the threshold of settlement up through the eighteenth

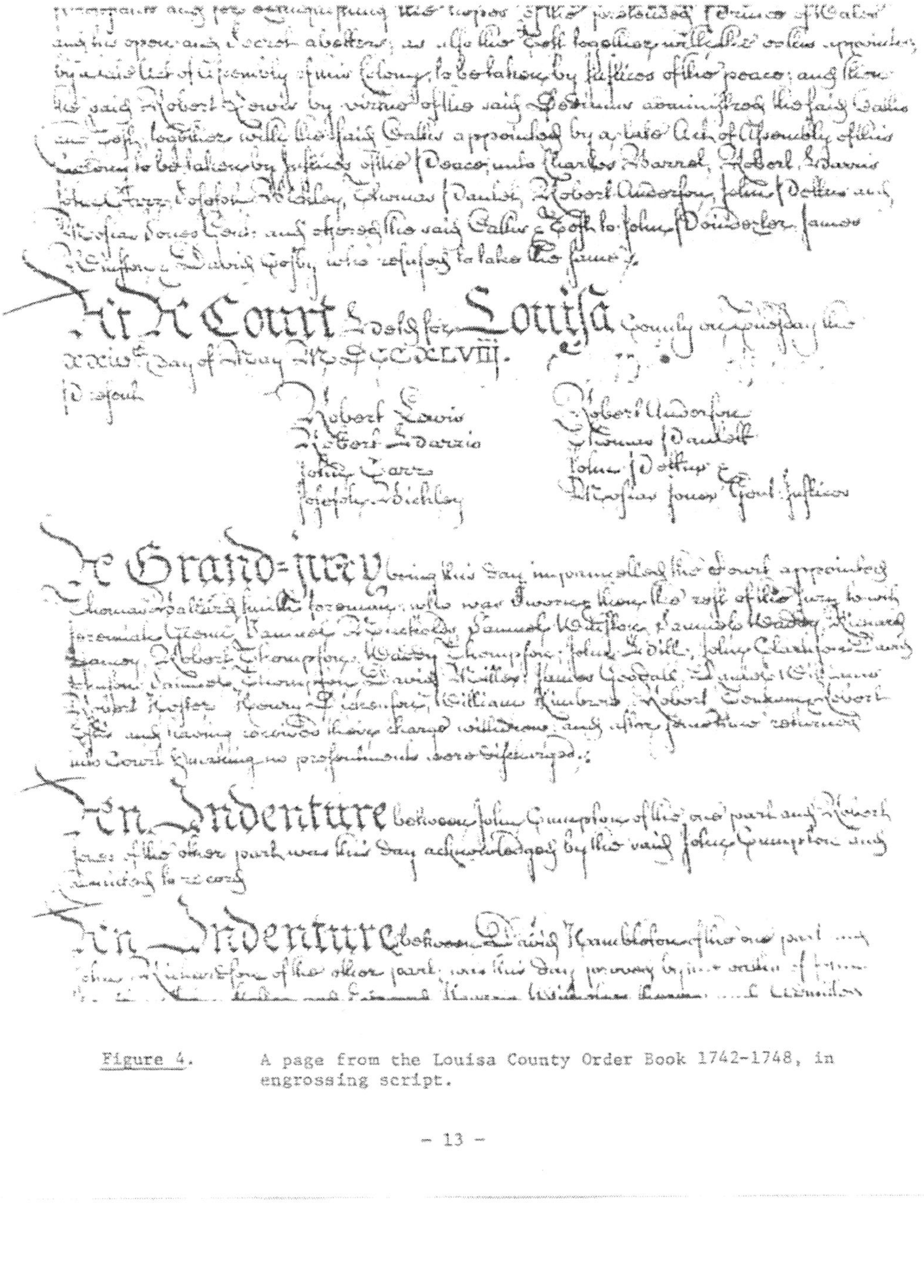

At a Court held for Louisa County on Tuesday the xxiv^th Day of May MDCCXLVIII.

Present

Robert Lewis
Robert Harris
John Carr
Joseph Bickley
Robert Anderson
Thomas Paulett
John Potter
Thomas Jones Gent. Justices

The Grand-Jury being this day impannelled [illegible]

An Indenture between John Thompson of the one part and [illegible] of the other part was this day acknowledged by the said John Thompson and [illegible] to record

An Indenture between [illegible]

Figure 4. A page from the Louisa County Order Book 1742-1748, in engrossing script.

century. Immediate corroboration for these chains of early road orders was provided by the fact that many of the principal roads studied had survived in all or part, under their eighteenth century name down to the present or the time when the Confederate Engineers produced the first good maps of some of the counties of Virginia in 1863 and 1864. Most of these early roads remain today nearly exactly on the routes shown on these maps. The ready availability of the road orders in a convenient printed form, along with these maps as corroborative evidence, greatly facilitated the writing of brief histories of some individual roads (such as Three Notch'd). These histories ultimately made possible the evolution of a road-oriented architectural survey course at the University of Virginia.

While the previously mentioned Three Notch'd Road report was being prepared for Mr. Bear at Monticello, K. Edward Lay, Assistant Dean of the School of Architecture at the University of Virginia,became aware of the road history project during an informal discussion with Howard Newlon, who in addition to his duties with the Council is also a lecturer in that School. His curiosity piqued by Newlon's descriptions of some of the taverns and other surviving buildings along the road, Lay, after viewing the slides, maps, and other materials of Messrs. Pawlett and Newlon, decided the potential existed for a very interesting, unique and innovative course tying architectural development to the existence of the road itself and its subsequent influence on Virginia history.

After discussions between the various parties involved, a course called "Architectural Patterns Associated with Eighteenth Century Virginia Road Traces" was initiated in late 1975 or early 1976 by the Architectural History Department. The initial semester of this course dealt with the Three Notch'd Road between the Blue Ridge Mountains and the Goochland County line. Following the success of this maiden voyage, it was decided that the course would be repeated. The second time, portions of the Secretary's Road and the Buck Mountain Road in Albemarle and Fluvanna Counties were dealt with. The documents from these surveys, consisting of information sheets, photographs, maps and brief histories, were bound and placed on deposit in the Fiske Kimball Fine Arts Library at the University, where they are available to scholars, local historians and genealogists.

In each case the availability of the printed and indexed road orders allowed the construction of a chronological chain of road orders for the road in question and, subsequently, a brief written history of the road supported by plats, deeds, and the corpus of the work dealing with the architectural survey itself. At the back of the published report on the Three Notch'd Road can be found an appendix containing just such a chain of road orders for it to which have been added other pertinent road orders from Orange and Augusta Counties, an ordinary license from Louisa containing the name, and citations of the name from the <u>Virginia Gazette</u> of Williamsburg.

The same system of recording used for the road orders will also be useful when extracting other information from early records, and will eliminate that curse of the historian and genealogist, longhand copying and notetaking while simultaneously allowing a literal reproduction of the documents in question.

For those documents such as plats (see Figures 5 and 6) and maps which must be directly reproduced from the original, either photography or a copying machine should be used. The great number and condition of the surviving plats in the Albemarle County Surveyors Books 1744-1853 showing roads made necessary the use of photography there, but it might in many cases be possible to produce usable reproductions on a copying machine where the number involved is not large. This assumes, of course, that plats of other counties contain as many roads as did Albemarle's, an assumption not generally supported by the plats of the several other counties which the author has thus far viewed.

Because the Albemarle plats did contain so many roads, an effort was made on each of the photographic copies to identify the present county and general area into which it would fall. Many of the plats fell within that period when Albemarle contained within it a number of other counties and the usefulness of this information to researchers over a broad area soon became obvious. Then, utilizing a form similar to that of E.G. Swem's Virginia Historical Index, an index was prepared and published as another part of the continuing series "Historic Roads of Virginia".

Many other sources of information already exist in a prerecorded readily available form. County road maps, for instance, can be procured from the Virginia Department of Highways and Transportation. Aerial photographs as well as detailed topographical maps, both by the U.S. Geological Survey, also come readily to mind, and are available at the Division of Mineral Resources in Charlottesville. Collections of early photographs similar to those by the Cooks in Richmond and Holsinger's in Charlottesville are also available in many places. Others are still held by individuals and are waiting to be discovered.

With the first really detailed maps of Virginia counties dating from the 1860's, and surviving photographs seldom of earlier date, a large gap of a hundred years or so exists between the coming into existence of most roads and their first appearance on maps of an area. Here enters what can be the most potent of all our research tools: the site survey. This is usually accomplished by actually driving over those portions of the old roads still in service on, or very nearly on, their original routes. Where this is not possible because of abandonment or alteration of the route, the old trace can usually be covered on foot in the fall or winter with only minimal difficulty.

At this time photographs of interesting surviving portions of the road or its trace, the architecture along it, and fords and bridges can be made and the road sketched on county road maps, or geological survey maps if necessary. Of necessity, the site survey usually will come only after considerable preparatory work has been done in the early road orders, plats and maps to establish the general route of the road in question and its degree of correspondence with the modern highway system.

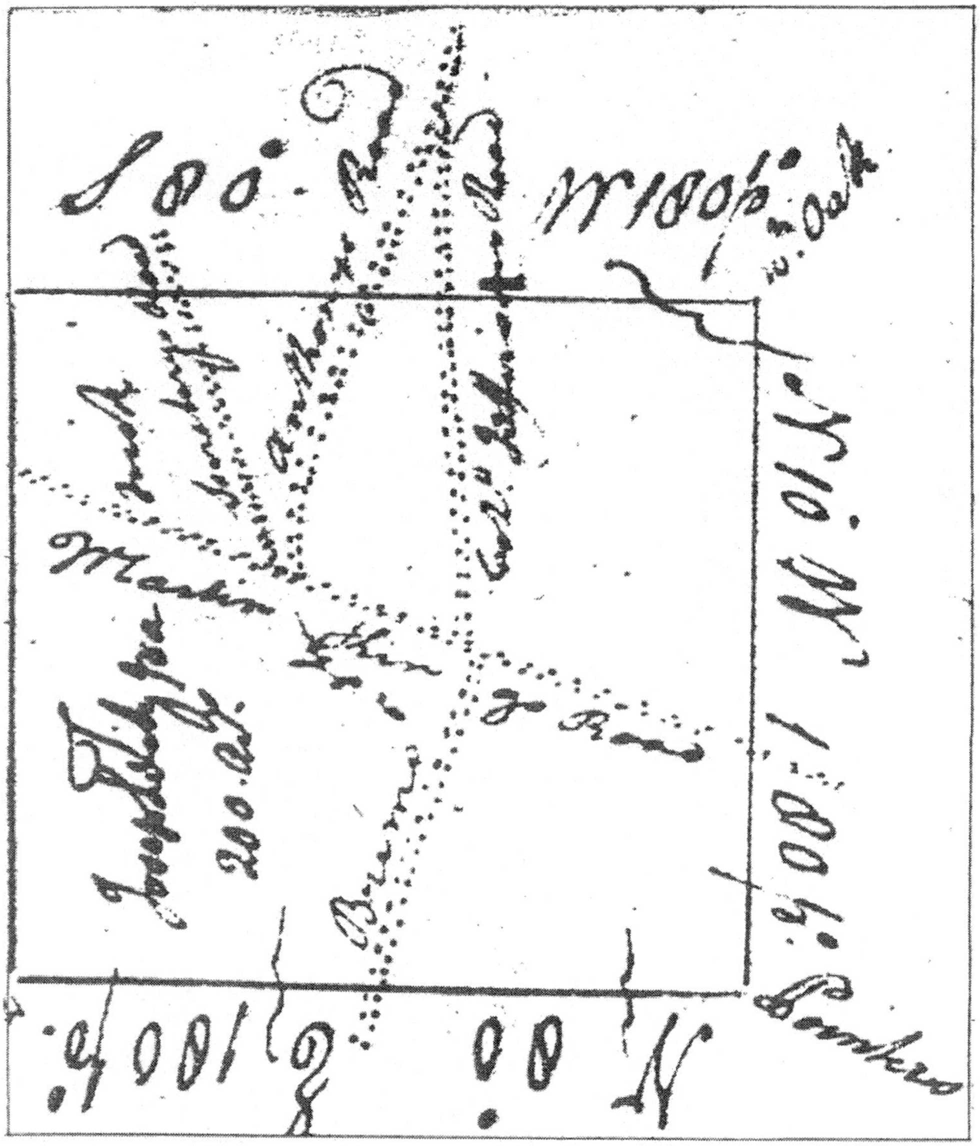

Figure 5. An unusual early plat (1755) showing the intersection of five roads at Woodridge in Albemarle County.

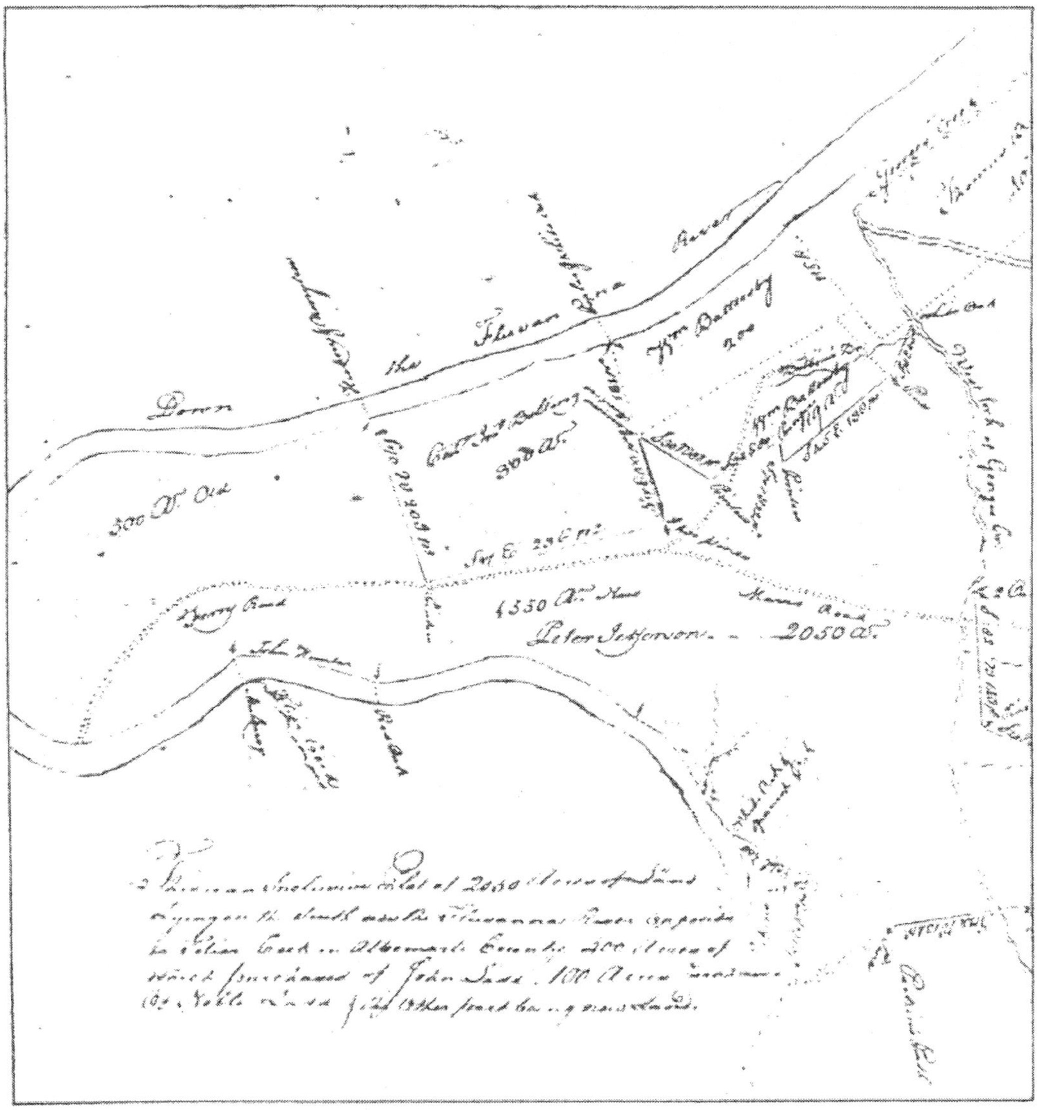

Figure 6. Another choice early plat, from the road historian's point of view, showing the Ferry Road and Marr's Road opposite Scottsville in Buckingham County.

After the novice road historian has done site surveys for several roads the logic of their routes will begin to become apparent to him, and the task will become progressively easier as he realises the constraints under which these early road builders acted and begins to appreciate their ingenious solutions to the problems presented to them by Virginia's geography. Although without modern technology, these men built roads the routes of which could be little improved by the later engineers responsible for the railroads and our present highway system. Of particular interest to the road historian is the scale of operations during the construction of some of these. Too often, a casual reading of a road order reveals only the names of several men and one fails to take into account the potential labour force available for road work when two or three planters of stature were assigned a road to clear or open. From the number of slaves known to be in the possession of many of these men at death, it would appear that the employment of 100 or more men at one time in constructing or opening a road would have been neither difficult nor unusual in the eighteenth and nineteenth centuries.

Besides the magnitude of the labour force employed, the amount of grading or levelling of hills is sometimes surprising in light of the notion prevalent today that such activities were extremely minimal in most of the early road building due to the lack of any sort of equipment beyond the pick and shovel of the individual labourer. What is perhaps forgotten is that often a lack of equipment can be overcome by a sufficiency of manpower properly applied, a fact still well known to people in a good part of the world. To actually traverse an eighteenth century road after a thorough study of the route on a topographical map is to convince oneself of the logic of the route and of the genius of those who laid out the road, whether animals, Indians, colonial Virginians, or all three. Time after time the budding road historian will see that the best, or the "least worst ", route was chosen by these men. Occasionally an inconsistency will seem to appear, but, invariably, a closer examination will reveal that this or that strange curve was designed to circumvent a swamp that no longer exists or to head or ford a stream at some shallower point.

One of the best and most readily accessible examples of the logic of a chosen route to which the aspiring road historian can be referred for study is the Three Notch'd Road. With the aid of the published study and map, and perhaps a few topographical or geological survey maps, this road can easily be driven from Richmond to the Blue Ridge in a day's time and the terrain, route and the structures along it closely studied. Any doubts concerning the routing can be dispelled by the student himself attempting to construct an alternate route as short and with as good grade, stream and river crossings, drainage, etc.

Besides the benefits derived by the student from repeating an extended site survey of the type represented by the Three Notch'd study, actually driving over a known eighteenth century road "up the country", the brief history of the early development of this road contained in the publication should provide an example similar to what will be encountered when the roads in other parts of Virginia are investigated. Combined with a guided tour of the road itself, this experience should enable the beginning road historian to start fitting together the various bits and pieces of information to be gathered from road orders, maps and plats in such a manner as to illustrate the development of the principal roads of an area while identifying them by name and defining their routes.

Evidence and Its Analysis

Road orders, the primary source of information, and their use in the construction of chains of evidence for such roads as the Three Notch'd have already been touched upon here. Deeds were not originally consulted in the Albemarle study, although it was supposed that they would prove very helpful for the early period if they were available in the form of an easily read, indexed typescript. The recent publication by Rosalie Edith Davis of Louisa County Deed Books A and B 1742-1759 seems to refute this idea, however, at least for this portion of Albemarle during the early period. When these deed books are placed beside a computer printout of all the place names contained in Louisa County deeds 1765-1815, prepared by Ransom True, nearly complete coverage is available for the first seventy-five years of Louisa deeds and the fallacy inherent in this line of thought is laid bare, at least as far as roads in the first twenty or thirty years after the initial settlement are concerned. Only nine roads are specifically named by the Davis index, with the rather unspecific "Main Road" leading in number of citations. Three Notch'd Road, the southern border of the county, is mentioned by name only once. It would appear that the earlier Hanover deeds for this area would be little more productive for this period, were they not lost. True's list, while containing more roads, is for a period beginning some forty or forty-five years after Louisa began to be settled, and thirty years after that part of Louisa later to be incorporated into Albemarle began to be settled. No doubt the problem arises from attempting to read the twentieth century practice of using roads for property lines back into an eighteenth century Virginia context where the original land patents were laid out with regard to the topography and the quality of the land (as well as other property owners' lines), rather than to road locations. In the beginning, of course, there were no roads to speak of, and those which did exist doubtless moved at the will of the users. It would, therefore, have been foolhardy to rest property ownership on the locations of such nebulous and transitory things as roads. Later, as the gentlemen justices of the county court made their power felt through their appointed surveyors of roads, locations became more nearly fixed, and as time and death took their toll and tracts changed hands and were broken up, roads probably became more important as boundary lines. While it is true that a considerable body of material has survived in the form of the later deeds and plats, these probably serve better as corroborative material than as anything else. Deeds and plats seldom can provide the answers to such questions as why a road was built, or when and by whom. To begin to answer these questions the original road orders themselves are essential.

Besides the historical limitations inherent in the use of the deeds and plats, there is also the problem of information retrieval, unless printed deed abstracts are available. While early plats can be readily examined, for an individual to extract from the early deeds of a county all the bits of information relevant to road development would be a very time-consuming procedure and still might ultimately prove of little benefit to the research. Where deed abstracts are available, as in the case of Louisa County 1742-1774, they may occasionally be resorted to with some benefit by the researcher, bearing in mind, of course, the rather limited use of roads as property boundaries in the early period.

Probably the most advantageous and effective use of deed information can be made by the spot search technique, which is essentially corroborative in nature. Here the area to be examined has usually been narrowed to several adjacent pieces of property thought to be the approximate location of an early road whose location (and name) may have been changed over a long period of time. Even then the title search and plat examination often will produce conflicting data and pose questions which simply cannot be resolved with the existing information.

A word might be inserted here concerning the dating of evidence as it relates to the initial opening of a road. With no good early maps existing for most counties, contemporary road orders are usually the best source, when they are obtainable, with plats, maps, deeds and whatever else is available serving to confirm the road orders. The nearer the date of any of these latter to the date of the initial series of orders concerning the road, the better they should be considered as evidence in a particular case. Now, as it often transpires that this other evidence may be separated from the orders by a considerable period of time, and that this evidence may conflict either with the earlier evidence, or even with all of the other later supporting evidence, it seems that some exclusionary rule should be formulated for such cases.

With this need in view, and taking into consideration the experience of all those associated with this project in attempting to research the eighteenth century roads of Albemarle, it would seem that the greatest weight should be given to that evidence falling within a 25-to 30-year period following the initial orders concerning the road, and somewhat less to that falling within a 30-to 60-year period after the initial orders.

That evidence dating as much as, or more than, 50 or 60 years after a road was initially opened should probably be viewed very critically. There are a number of reasons for this. First, in the eighteenth century a period of this length would probably have exceeded the lifespan of most men, and therefore of living memory, and would have become dependent to a great extent upon traditional knowledge, that as handed down from father to son. Thus, errors might have crept in during the second generation. Second, roads constructed about the threshold of settlement and during the succeeding two decades or so were probably more likely to be moved slightly during this time as deficiencies existing in the original route came to be corrected. Then, as population increased and new counties were created, county seats moved, and boundaries rearranged, routes of travel were often altered and what had been an important road turned into little more than a disused path.

Beyond this, the historical forces external to the various counties but acting upon them often have to be taken into account as the tide of settlement rolled westward across and through the Valley and into the present states of West Virginia and Kentucky. This settlement probably began to make itself felt more in the years following the Revolution when legislative enactments of specific road-related nature began to occur rather frequently. Sometimes a researcher working backward from a late eighteenth or early nineteenth century piece of turnpike or bridge legislation will be able to attach new significance to certain road orders issued by the county court or that of an adjacent county, and will find that all of these precursors of the turnpike or bridge related to the developing transportation needs of some distant county or area of Virginia or West Virginia, or perhaps even Kentucky. Often, with the benefit of a knowledge of later events, what at first appeared a rather cryptic road order will take on new meaning, sometimes explicitly stating what was only imperfectly understood before by the researcher. Many cases of this sort should come to light as the records of the Board of Public Works 1816-1902 are examined by individuals working on turnpike and road studies.

THE SCOPE OF A ROAD HISTORY

After one has surveyed and mastered the various techniques of the road historian and the collection of information is well under way, the question ultimately must be asked: What will constitute a satisfactory road history for the particular area under consideration? Although the answer will to some extent depend upon the area in question, its history and state of development, and its surviving records, it would be well to here attempt some sort of a very general definition of what would constitute an acceptable road history.

Given that the art is presently in its infancy, it would first seem wise to then place some limits on the coverage any road history of a county would attempt at this time. A multitude of reasons could be adduced for this, but the two principal ones should suffice.

First, the Albemarle experience, if taken as typical, would indicate that most of the principal road network of a county has usually been laid down within twenty-five years of the initial settlement of the area, and will have survived intact, though not without some modification here and there over time, to the present. Beyond this, certain landmark events such as the American Revolution and the era of the turnpikes, canals and railroads will have their effects and will have to be taken into consideration, but it is doubtful if this rule will be greatly modified by future experience in the field. Where possible, the study should also encompass these events and their effect upon the roads of the county, but not at the expense of stretching the study to cover too long a period. The formative years are really of the greatest importance. After these have been thoroughly examined, studied and mastered, the subsequent years can be studied by others. Turnpikes, canals and railroads can themselves each be made the subject of separate studies which may relate in whole or in part to the study of the early roads.

Second, also taking into consideration the Albemarle experience is the relative difficulty in coping with the mass of repetitive road orders should a period of, say, a hundred years or more be studied at first. With the sub-division of roads and the resulting proliferation of surveyors the task becomes well-nigh impossible, at the least much too time-consuming for its productiveness of new information, and coverage of this period might best be deferred to a later time, and perhaps also to different techniques.

From the foregoing it would then seem that beyond a certain point the real return on the time spent in research would decrease. Therefore, it is suggested that the period to be covered by the first road history of a county be limited to no more than, say, fifty years from the threshold of settlement of the area in question. Notice that the time span here is taken from the period of first settlement rather than from the date of formation of the county.

It might even be possible to gain an adequate picture of the early roac of a county by covering a smaller period. It is probably better, however, t err in the direction of covering too great a time span rather than too small a one. Here, rather than being able to make a positive statement based on solid experience, the author must indulge his feelings. While the Albemarle study in theory spanned some 86 years, records' destruction covering about 35 years (1748-1783) effectively reduced the period to near the half-century mark. With this blank space occurring just in the middle of the study, the author can only say that he thinks, based on his experience, that the fifty-year period from about 1730 to 1780 would have covered most of the really significant road building activity in Albemarle County. He also feels this will prove to be the case with the roads of other areas in Virginia when studies are conducted there.

A period of fifty years would also seem to be well within the realm of possibility when the interlocking nature of road studies is considered, the fact that as more of these studies are completed researchers beginning new histories will find significant portions of their work already completed for them, the road orders published, etc. because of overlaps in coverage, subsequent subdivisions of counties, or the fact that adjacent counties will already have published road histories available. Each study completed will to some extent serve as seed for the next until the massive picture puzzle of Virginia's early roads will again lie before us completely reassembled.

Road Order Coverage

The limitation of 'coverage to a finite period calculated to cover most of the early road development will also serve to make possible the handling, collation and indexing for publication of the road orders of a county without excessive effort and expense. The overlapping nature of the whole road history effort should preclude the necessity for publishing as much as fifty years of road orders for all areas, once the first few, pioneering studies have been completed and published.

The usefulness of published road orders extends far beyond the history of the roads themselves, into the fields of local history, genealogy, architectural history, sociology, folklore and others still to be enumerated, as new uses continue to be found for them. Road orders are particularly important during the quarter century or so following the initial settlement when the usual paucity of surviving local records makes it so difficult to see exactly what was happening, who was there in residence in the area, and what the problems of trade were.

Plat Coverage

Early plat coverage will probably not be available for all counties; perhaps not for most of them. Even where the early surveyors' books have survived it is unlikely that road coverage will equal that of the Albemarle books. Where this does happen an index similar to that done for Albemarle would seem to be a worthy addition to the project and be productive of long-term benefit.

Where the number of plats showing named roads is not so great these might simply be cited in an appendix to the road history itself or to the published road orders rather than be given separate publication.

Site Surveys

Site surveys are the core of any adequate road history, the sine qua non so to speak, tying together the information gathered from road orders, plats, and other sources, verifying the conclusions drawn from this information by the researcher and, often, providing fresh flashes of insight as the road historian attempts to place himself in the position of the eighteenth century surveyor laying out a road along the best (or sometimes, the "least worst") route through the countryside of Virginia.

It is doubtful whether any road history worthy of the name could be produced without a fairly comprehensive survey of all of the main roads of the area, their geography, river and stream crossings, and so forth. Certainly this is true for the early principal roads of each area, although probably somewhat less true of the later, subsidiary, connecting roads. By performing the surveys faithfully the researcher will shortly begin to develop a real "feel" for the early roads. This will enable him to speak with much greater authority when he puts pen to paper than would one who remained in a cloistered study populated only with books and maps, or in the county courthouse poring over the records, and attempted a road history from there.

The results of the site surveys should, however, be used as confirming Or supporting evidence in writing the road history, rather than being incorporated simply as narratives of trips along certain roads with a point-by-point commentary 011 what was seen. This should not preclude the citing of significant facts about the present condition of an old route or its surviving trace alongside a more modern route, its fords and grades.

Illustrations

Quantity and placement of illustrations, whether photographs, prints or line drawings produced especially for the road history, will have to be determined by what is available for the area in question. Since there is probably a finite number of existing period prints and photographs of a general nature (i.e. tobacco hogshead being rolled, stagecoach travel, etc.) repetition may become a problem after a few histories have been published using many of the available ones. Later histories will probably have to lean more heavily on illustrations relating to their specific area, to modern or earlier photographs and/or sketches.

Naturally, the visual materials available for some areas will be far greater than those for others. Where no photographic archives exist researchers will be forced to fall back upon their own ingenuity and resources. Moder photographs, conjectural drawings and panoramas all can be used to illustrate key points in an early road network, or even the whole network. To be avoided at all costs is the useless repetition of those prints already overly used to illustrate books and articles on Virginia's early history.

<u>Narrative</u>

What form should the road history itself take as it is written? No really comprehensive answer to this question can be made, although it is assumed that the form will be generally that of a rather straightforward chronological narrative as it unfolds itself to the reader.

Variations in the patterns of development of the various counties, differences in the kinds of surviving records, architectural and road remains, and various other factors will doubtless serve to greaten the emphasis on certain parts of the narrative and to lessen it on others. Certain aspects of the history may also call for extended and detailed treatment as they occur in the narrative. As with the Albemarle road history, this will usually be determined by the first occurrence or occurrences of something such as, say, bridge building or river navigation, when it will usually be found most convenient to insert an essay dealing with the subject in some detail for the whole early period, or at least up to some landmark or watershed event in the field. In cases where this insert would be of such magnitude as to render the narrative discontinuous it would perhaps be better to consign the information to an appendix while allowing a brief synopsis to carryon the narrative to its conclusion.

<u>Maps</u>

As with the matter of illustrations, the choice of maps for a road history will depend greatly upon the ingenuity of the individual researcher. The term "map" is here used to define those maps detachable from the printed study itself, rather than those maps or portions of them contained within the study and used to illustrate portions of the narrative.

One of the main problems inherent to preparing a road history is that of making the narrative understandable to the novice reader in terms of the present landmarks, roads, route numbers, towns and villages. This becomes even more difficult if adequate maps are not supplied with the narrative. Detachability is specified because it is anticipated that almost all readers will, at one time or another, wish to conduct their own site surveys of various roads or portions of them dealt with in the text.

While theoretically it would be possible for one interested in road history to follow the history of a county's early roads using only a current road map of the county, a much clearer picture is possible if, in addition to a current road map, a map drawn on a similar scale and showing the early road network is at hand. In the interests of illustrating the apparent correspondence of the present road system with that laid down within the first quarter century or so after the threshold of settlement, this may take the form of a semi-transparent overlay. For this purpose it should not be necessary to show every early road, but rather the principal roads of the time. Indeed, to show more would probably only serve to further obscure the view of the reader.

Maps, or portions of them, used to illustrate either specific portions of text or specific areas of the county may be taken from historical maps such as those by the Confederate Engineers, or from plats, or they may be sketched to the particular purpose at hand. These, however, will then become an integral part of the text and are probably best considered under the heading of illustrations.

In some situations it also may be found useful to prepare a detailed set of geological survey maps as a part of the project. These would be very useful in areas undergoing rapid urbanisation where evidence of the early roads is being obliterated or is in danger of being so in the near future. These maps would not be published but rather would be placed upon completion in some convenient repository. Similarly prepared examples of this type of map are on deposit at the Fiske Kimball Fine Arts Library at the University of Virginia in the previously noted volumes on architectural surveys along early Virginia road traces.

SUMMARY

The preceding points relating to what a road history should contain might be reduced to the following:

1. Fifty years or less would seem to be the optimum length for the first study of the roads of an area.

2. Road orders should be published in an indexed volume or series of them covering this period or that portion of it not already covered in published volumes of road orders.

3. Plats showing roads, if present in large numbers, should have indexes prepared and published. Any other information of general applicability developed during research should receive similar treatment so as to make it easily retrievable.

4. Site surveys are designed to help the researcher to interpret th archival and cartographic information developed from state and loca records. Records or notes of site surveys, if kept, might be store with the detailed geological survey maps (also optional) at som public repository.

5. The matter of illustration will in many, if not most, cases allow, perhaps even require, innovation if undue repetition is to be avoided. Even if new sources of original prints and photographs are unearthed it will probably soon become necessary to produce for each work some (at least) original illustrations. These could be similar to those found in Convenient Wayes.

Figure 7. The 1864 Confederate Engineers' Map of Albemarle County.

6. A road history should be written as a straightforward chronological narrative, with brief, occasional digressions to cover special topics.

7. Each road history ought to be equipped with at least two detachable maps. One of these should be a current county road map; the other one should be a specially constructed map showing the early road network and related important historical sites. The scale of both maps should be the same and the latter map should be sufficiently transparent to be used as an overlay on the former.

The final, most important consideration is that all of the above combine to produce an end product such as to enable the layman, be he amateur historian or casual reader, to readily follow the narrative using the enclosed road maps. All else goes for naught if this should be lost sight of in writing a road history.

The reader should, with a minimum of difficulty, be able to understand the development of the early road system from a careful reading of the narrative. Ideally, he will also be able to use these road descriptions and the maps to drive over most of this system and see for himself the ingenuity and wisdom of the early road builders.

INDEX

Printed in Dunstable, United Kingdom

91222383R00020